DAYS of the YEAR

CHRISTMAS

MAN AND RELIGION SERIES
Part 2
LIVING IN BRITAIN

DAYS of the YEAR

JOYCE McLELLAN

THE RELIGIOUS EDUCATION PRESS
a member of the Pergamon Group
Headington Hill Hall
OXFORD

JOYCE McLELLAN
is
Senior Lecturer in Education
Bishop Otter College of Education
Chichester

Illustrations
MEG PATERSON

First published 1971

ACKNOWLEDGEMENTS

The author wishes to thank the following for permission to use these extracts:
Cassell and Co. Ltd., for the carol in chapter 16,
THE TWELVE DAYS OF CHRISTMAS
from
THE TWELVE DAYS OF CHRISTMAS
by Miles and John Hadfield.

The Clarendon Press for the two rhymes in chapters 7 and 10 from THE LORE AND LANGUAGE OF SCHOOL CHILDREN by Iona and Peter Opie.

The Oxford University Press for the following carols from the OXFORD BOOK OF CAROLS:
in chapter 8, The World Itself keeps Easter Day;
in chapter 10, May-Day Garland
(collected by Geoffrey and Martin Shaw);
in chapter 16, The Holly and the Ivy.

MAN AND RELIGION SERIES
General Editor Ronald Dingwall
Executive Editor Gordon Hawes
Art Director Keith Clements

THE RELIGIOUS EDUCATION PRESS
a member of the Pergamon Group
OXFORD NEW YORK TORONTO
SYDNEY BRAUNSCHWEIG

Made and printed in Great Britain by
A. Wheaton and Company
Exeter

L.O.C. 78–109235

08 006799 9 (hard cover)
08 016278 9 (flexicover)

CONTENTS

1 INTRODUCTION

Thirty days hath September,
April, June and November,
All the rest have thirty-one
Excepting February alone,
Which has but twenty-eight days clear
And twenty-nine in each leap year.

This is a rhyme which we use to help us to remember how many days are in each month. How do you remember special days? Have you a list of special days in the year? These might be days when you have presents, like birthdays, or days when you wear special clothes and eat certain kinds of food.

In this country we have many days which stand out among all the ordinary days in the month. Some of these are known and kept by almost everyone, such as Christmas Day, and some are known mostly to the country people, like St. Swithun's Day. Some of these days are solemn, religious days, like Easter Day, and some are funny ones, when everyone plays tricks, like 1st April which is known as All Fools' Day.

Some days have special songs, poems and dances, like May Day. Other days have something to do with happenings in history; one of these is 5th November, Guy Fawkes' Day. There are the days which are called after people who lived hundreds of years ago and about whom many stories have been told. The 15th July, St. Swithun's Day, is one such day.
There are also those days which have strange customs so old that no one knows when they were really started. The 31st December, New Year's Eve, and 31st October, All Hallows' Eve, are two of these days.

Other days are important to us because they have something to do with the four seasons in our year—spring, summer, autumn and winter. Many of our customs are connected with the weather of the four seasons. Have you noticed that we talk about the weather in this country? We say "It's a lovely day today" or "It's better than it was yesterday". Why is the weather so important to us?

Before we could buy food from other countries and have it sent to us by ships and planes, we had to grow most of it ourselves. This was hundreds of years ago, when our people lived mainly in small towns or villages and not in large cities as they do today. Crops were grown, such as wheat, oats and barley. Apple and pear trees provided much of the fruit which was eaten and wild fruits like blackberries and rose-hips were carefully gathered and stored as jam. Hazel nuts were also stored. All these things need rain to make them grow and sunshine to ripen them. Therefore the weather was very important to the British people. Too much sun or too much rain at the wrong time could mean that the crops would wither or that the people would be unable to gather them at the right time and then they would have a hard winter.

Before people knew the right way to grow crops and fruit, they thought that success in these things was a matter of luck or chance. A number of customs grew up to make sure that the luck would be "good luck" and would give them enough food for the coming year. It was quite hard work to do all the things that would bring this good luck, but the people would not leave any of them undone for fear of what might happen to them in the long hard winter.

Sometimes the people thought that fairies and witches had something to do with the good and bad things that befell them. These "fairies" were not the kind which we sometimes see in pictures with wings and magic wands. They were the "little people" who lived in trees, under stones, in rivers and wells and other such places. The poet William Allingham tells us about them in his poem called *The Fairies.*

Up the airy mountain,
Down the rushy glen,
We daren't go a-hunting
For fear of little men;
Wee folk, good folk,
Trooping altogether;
Green jacket, red cap,
And white owl's feather.

Down along the rocky shore
Some make their home,
They live on crispy pancakes
Of yellow tide-foam;
Some in the reeds
Of the black mountain lake,
With frogs for their watch-dogs
All night awake.

These fairies could be kind and helpful and the people then wanted to reward them for their kindness and would leave presents for them. Sometimes they played mischievous pranks and then they had to be given presents to make them good tempered. We shall see how some of the things people do on New Year's Eve are concerned with keeping "on the right side" of the fairies at the beginning of a new year.

When Christianity was brought to this country, hundreds of years ago, by bishops from other countries, they tried to change some of these old customs of the people. They helped the people to celebrate some of the great days of the Christian year, such as Easter Day and Christmas Day. But the simple country people kept some of their old customs and used them as well as the religious ones, so that some of our special days have a curious mixture of old folk-lore and religious ceremonies.

On the front of a British Passport you will see that our country is called the United Kingdom of Great Britain and Northern Ireland. This United Kingdom consists of England, Scotland, and

Wales as well as Northern Ireland. Our flag, the "Union Jack", is designed from the flags of three of these countries. They are placed in such a way that if you look carefully each separate flag can be seen, while giving the effect of one single flag.

The national flowers of these countries can be seen on some of our coins and postage stamps. Until the last issue of these bearing the Queen's head only, somewhere on the most common stamps it was possible to see the rose of England, the thistle of Scotland, the shamrock of Ireland and the daffodil of Wales.

Each country also has a patron saint. These were Christian people, sometimes bishops, who had a particular place in the history of the country and about whom the people liked to talk and think. They often felt that these saints protected their country and many stories grew up about wonderful miracles which they performed. One saint had so many legends surrounding him, that one writer said that while the saint's name was quite rightly well-known to all the people, the actual things which he had done were known only to God!

The four patron saints of the United Kingdom are St. George of England, St. Andrew of Scotland, St. Patrick of Ireland and St. David of Wales. These saints each have their own days in the year. These days are remembered in their countries or by the people of that country wherever they happen to be. Their own flags are flown and people wear the flowers of the country. In the chapters about these days some of the most well-known stories are told. There are others which can be found in books of legends, but although they are interesting, they are sometimes very fanciful and difficult to believe.

In the following chapters you will find accounts of our special days. You may be able to add other things to the ceremonies which are described here, and at the end is a list of books to help you more.

2
NEW YEAR'S EVE

It may seem strange to start a book about the days of the year with the very last day of all, 31st December, but long ago people believed that the last hours of the old year were the proper time to put away all the bad habits of the old year and to find good ones for the new year. They had rhymes to be said and certain things which had to be done before twelve o'clock at night.

Ring out the old,
Ring in the new,
Ring out the false,
Ring in the true.

This was one of the rhymes set to the peal of bells which rang in the churches to herald the new year. Now, as well as the bells at midnight, many churches have "Watch Night" services; at Basingstoke, a hymn is sung by the choir in the church tower so that it can be heard by many more people than those who are joining in the service inside the church.

In Scotland and the North of England many very ancient ceremonies are performed, which date from the days before Christianity came to Great Britain and the people believed in fairies, witches and spells, and in the powers of the trees.

In Scotland New Year's Eve is called "Hogmanay". This is a very old word and no one is sure what it means. Some think it may have something to do with the fairies. Certainly some of the old acts which are performed on New Year's Eve had some connection with them. Some housewives in Scotland would clean their houses from attic to cellar very thoroughly on this day. It was considered that good fairies preferred clean houses to dirty ones and were more likely to visit them and to bring good luck. The wife would take great care to remove all the old ashes from the fireplace for fairies were very offended if the fire was not new and bright.

Some food had also to be provided, perhaps a kind of cake or a fruit bun and also a bowl of cream could be put down by the front door. While all these preparations were going on there were various rhymes which the wife could say or sing which would act as extra safeguards against any of the pranks which the fairies might choose to play.

Although she could do many things to put the fairies in a good mood, there was one thing which she was warned never to do. This was to leave a suit of clothes for these unseen visitors. Food they liked but the gift of garments would make them feel that they owed something to the members of the household. They would then feel as if they had become the slaves of human beings and would be angry.

Some of these ideas may sound strange and even silly today, but they have become the ritual of New Year's Eve and are now enacted mainly for fun. For the Scots like to have a party at Hogmanay and eat and drink toasts to welcome the new year. One of the acts which was supposed to have been performed by the fairies has now been taken over by the human beings. This is "first footing". Sometimes people leave their own parties to go and visit another house in order to be the first "foot", or person, to cross the threshold or doorstep in the new year. The proper presents to take on this trip are a lump of coal and some salt. This is to make sure there will be fire and food in the house all through the year.

If a dark person comes "first footing" it is considered to be luckier than if the visitor is a fair person. Dark fairies were always thought to be luckier than the fair ones; perhaps this came from the idea that the "brownies" were the best fairies to welcome to the house as they were hard-working, kindly folk, who would rather be helpful than play tricks on their hosts. The kind of things these brownies did to help humans can be seen in many of the folk stories of Great Britain.

The old year is often shown as being a very old person and the new year as a young baby. So some people ask the oldest person in the party to go out of the back door and the youngest one to come in through the front door, in order to open the way for new and pleasant things in the new year.

Other people have adopted some of these old Scottish customs into their New Year's Eve parties, for this often finishes the round of parties begun on Christmas Day. After 1st January people seem to settle down into the new year and forget the festive season. There is usually a period of hard work before the days which commence the period of Lent and herald the first days of Spring.

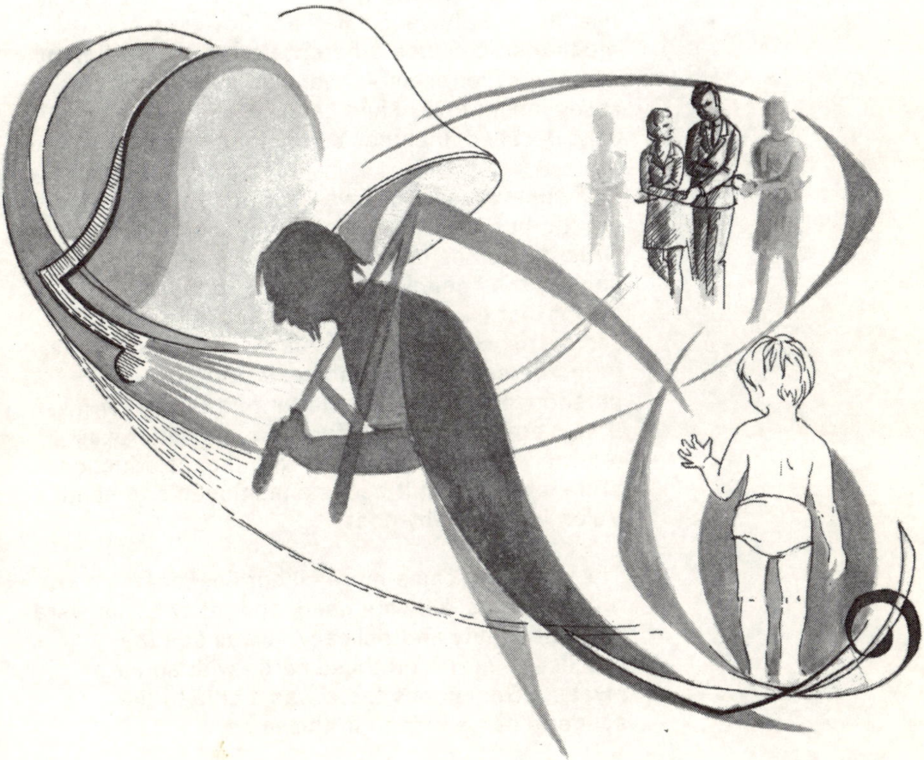

3
St. VALENTINE'S DAY

Roses are red,
Violets are blue,
Carnations are sweet
And so are you.

The 14th February is called St. Valentine's Day. Valentine was a priest who lived in Rome. He was martyred on 14th February for sheltering Christians instead of betraying them to the Romans to be persecuted. The middle of February was also celebrated because it is supposed to be the time when the birds choose their mates. Over the passing of the years these two events have become associated and Valentine has become known as the patron saint of lovers. On his day messages were sent to declare love and admiration. They were usually sent without a name, so that the recipient had to guess who the sender was. All kinds of verses have been written for this purpose, like the one above. Sometimes flowers were mentioned because different flowers had different meanings; for example, pansies meant thoughts, roses meant love. The "Language of Flowers", as it is called, includes a very long list.

Sometimes messages were carried by word of mouth, but eventually it became fashionable to write the verses on a card. In the nineteenth century the "penny post" was introduced. This meant that many people could afford to use this service. They could buy a stamp and send a letter to the person of their choice. Before this, the person who received the letter had to pay for it. If he were not able to afford this, the letter was returned to the sender. But with the introduction of the cheap post it became fashionable to send Valentine cards by post.

These then became more elaborate. Lace, ribbons and silks were used, and many cards were of great beauty and delicacy. It was still the practice to surround these cards with an air of mystery. Sometimes there was a clue to the sender's name hidden in the verse.

Many cards were made entirely by hand, with great skill involved. Lace was hand-made and fashioned with great care to form a base or mat for embroidered or pressed flowers. Later the lace borders were made by machine and the different publishers had characteristic patterns so that it was possible to identify those made by the same firm.

Some of the early cards had parts which moved by pulling a short string and revealed a second small picture inside the first. For example an opening was made in the side of a ship to show a picture of Cupid visiting a sailor, another similar design had a small picture of a sailor writing his Valentine poem. Another favourite design was the kind which stood away from the card when unfolded, and in this form, views of the countryside could look quite realistic.

A masterpiece was the kind of Valentine card called a "cobweb" or "beehive". This was made of two layers of card. The under one had a picture delicately painted and the top piece was very finely cut so that it was quite flat and covered the under one. By means of a string this could be raised so that it formed a cage or net through which one could see the inside.

These beautiful hand-made cards gradually gave place to machine-made ones. These were often made in different shapes of astonishing originality; some were like flowers themselves, or birds, or fans or cages with various creatures inside. Some were so large and covered with shells, dried flowers, grasses, beads or feathers that they needed large boxes to put them in, and these in turn were decorated.

It seems that some people had reached the end of their ideas for beautiful ones and so they turned to jokes. At first these were mild puns, but they became almost crude in the search for something new and different.

These cards are sent in great numbers today and it is possible to find examples of the sentimental kind with flowery pictures and verses and also the cruder kind with jokes and puns. No one, however, seems to have time to make the delightful kind which were popular just over a hundred years ago. It seems a pity that the skill of making the cobweb Valentines has been lost to our modern world.

SHROVE TUESDAY

Shrove Tuesday is known to most people as "Pancake Day". It is usually in the month of February but the actual Tuesday on which it falls depends on the date of Easter Day. At present Easter Day is fixed by the movement of the moon, but many people hope that we shall soon have a fixed date for Easter. Then of course, Pancake Day will have a fixed date also. It is forty-one days before Easter Day, the day immediately before the beginning of Lent (Ash Wednesday).

On this Tuesday, in the early days of Christianity, people were told to eat the fats, eggs and flour which they had in their houses in order to prepare for the great fast which took place in Lent. This is the period of forty days before Easter. Lent starts with the services of Ash Wednesday, so the day before this was called Shrove Tuesday. Today we no longer use the verb "to shrive" of which the word "shrove" is the past tense. It means to present oneself to a priest, to confess one's sins and wrong-doings and so to prepare oneself for God's blessing. On Shrove Tuesday, people had to prepare themselves, by visiting the priest in Church, and also to prepare their households by using any extra stores they might have, so that their souls and bodies would be ready for the long fast of Lent. During Lent they would discipline themselves to eat very little simple food. Sometimes this was only bread and water.

The simplest way of using their stores on Shrove Tuesday was to mix eggs and flour together, sometimes adding milk to make the mixture liquid and then to pour it into a frying-pan in which butter or some other fat had been heated. This mixture was cooked until it was brown on one side and then it was turned over and cooked on the other side. If the pan used were a large round or oval one, the commonest kind for large families, it was a very difficult task to turn the "pancake" over without breaking it. It was found best to toss it up in the air so that it turned itself over! Then the housewife had to catch the pancake in the pan before it fell to the floor. This is a very difficult thing to do, but the cooks practised until they were very skilled at tossing the pancake.

In some parts of the country there are races organised to see which cook is the quickest and most skilful in this art of making and tossing pancakes. Perhaps the most famous of these is at Olney in Buckinghamshire, where the first race was held in 1445. A church bell is rung just before noon. This is known as the pancake Bell. This race is only for women and they must all wear aprons and caps or something to cover their hair. They must toss the pancake three times as they run and the winner receives a prize of a prayer book from the vicar and a kiss from the bellringer.

In recent years the annual race has become international, with competitors in the town of Olney in Texas, in the United States of America, running over a similar distance, the winner being the first woman to complete the course in the shortest possible time.

Shrove Tuesday used to be a general holiday and besides eating pancakes, games and sports of all kinds were enjoyed. The Pancake Bell was the signal for the games to start and another bell was rung in the evening when they had to stop. There were games of football played in the streets and in earlier times wrestling, cockfighting and races took place.

Children would go from house to house and be given pancakes or cakes, sweets or money. If they did not receive anything, they might throw stones at the door! In a Derbyshire village the custom of barring the schoolmaster out of his school was carried out on Shrove Tuesday, which was always a school holiday.

There is not much evidence of the games and revels of Shrove Tuesday to be seen now, but the practice of making, tossing and eating pancakes is a strongly established habit and is observed in almost every home in Great Britain today.

5
St. DAVID'S DAY

The first day of March is St. David's Day. He is the patron saint of Wales but unfortunately little is known about him. Some stories remain which may or may not be true, but it is clear that he was much beloved by the Welsh people in the sixth century when he lived among them. He was also a very outstanding preacher. There are many stories about his having preached to his people and given them such encouragement and help that, when he died, they were overcome with grief and said that they had lost their father and they wondered who would now look after them as David had.

He is reported to have preached to thousands of people gathered at Llanddewi-Brefi in a valley in Cardiganshire. As well as David there were a great many bishops present who were worried because they feared that such a vast number of people would not be able to hear them when they spoke. David, however, was able to make himself heard, and preached on behalf of all the bishops of Wales. Some people believed the story that, as David preached, the ground under his feet rose up until it became a mountain. Afterwards a church was set up there and dedicated to St. David. There is a cathedral in Wales called St. David's which for many years was regarded as a shrine to the Saint and many pilgrims visited it. The cathedral still stands there but today is used for regular services rather than as a shrine for pilgrims.

The daffodil is the national flower or emblem of Wales. This can be seen on coins and stamps but the leek also has connections with St. David's Day. There is some confusion in the stories about the origin of the Welsh wearing the leek on St. David's Day. Some people say that St. David was called upon to help his people to fight against an intruding army. The place where the two forces met was a narrow valley and the opposing sides quickly became mixed up together and it was difficult to see which was friend and which was foe. St. David is reported to have said to his people "Pick the leeks which are growing nearby and stick them in your helmets". This they did and so distinguished themselves from the enemy.

The soldiers in Welsh regiments today are presented with leeks on 1st March. It is quite impossible, however, that St. David could have been present at the original battle as it happened some years after his death.

Another explanation of this association of St. David and leeks is that he is reported to have lived a very simple life eating only such things as bread and leeks, and because of the wonderful sermons he preached and his helpful encouragement of the people, they concluded that leeks must have some magical properties.

He is the only one of the four patron saints of Great Britain to have worked in and to be venerated by the people of his native land. Many churches in South Wales are dedicated to him and David is one of the most popular names given to Welsh boys.

St. PATRICK'S DAY

St. Patrick's Day is 17th March.

St. Patrick was born about AD 373 but it is not certain where he was born. Some records state that it was in Scotland and some say that it might have been in the Welsh border country. However, all the accounts seem to agree that his father and his grandfather were deacons in the Christian church so that Patrick knew something about Christianity from the time he was a small boy.

When he was about fifteen a band of raiders appeared at his home and carried him off to Ireland with them. He was kept for six years in the country in North East Ireland. In those days it was a very wild part, mostly boggy and often bitterly cold. His job was to mind the herds belonging to the chieftain of the area. He was treated as a slave and no one cared for him. In those lonely days his only comfort was to think about his early teaching from his father and grandfather and to hope that one day he would be able to return to his own home.

One night he had a dream. It was so vivid he felt that it must be a message from God. He saw a ship and heard a voice telling him that one day he would return to his own country. He took this to mean that he should escape from his master and try to find a ship to take him away from Ireland.

After many adventures, he at last reached the coast where he found a ship. This was full of wild Irish wolf-dogs which were to be sold to a foreign country. After some difficulty, he persuaded the captain to take him on board, and they sailed away from Ireland.

It is thought that Patrick landed somewhere in France and wandered in that country until he came to a monastery. Here he stayed and studied for some years and eventually he returned to his own home country. Even though he was glad to be with his own people at last, he could not settle down happily at home. In his dreams he constantly thought he was back in Ireland and that the people there were calling him to come and teach them about God and to found a church there.

So he returned to France in order to prepare himself for his mission to Ireland. He gathered a small band of people round him and they set off to teach and to preach there.

There are many stories told of the miracles which Patrick did, helping the people to forget their older pagan ideas and to worship God. One of these stories described the way in which Patrick was supposed to have got rid of all the snakes from the country. There were so many reptiles there that they were like a horrible plague, and the people were very frightened as their bites often proved fatal. One day, Patrick went to the top of a mountain. When he reached the top, he drew all the snakes up after him, by some tremendous power, and then drove them down the other side of the mountain until they reached the sea where they were drowned. It is said that to this day there are no snakes in Ireland.

There is another story which is much more likely to be true. Patrick found the people simple and not very easy to teach. They found it hard to understand the important facts of Christianity, and in particular they did not understand the doctrine of the Trinity, that there is one God but also three persons in the one God. So Patrick picked a leaf of a little green plant growing by the wayside. He said, "This little plant, the shamrock, has three parts to each leaf but these three parts make one whole leaf. So it is with God, there are three in one, God the Father, God the Son and God the Holy Spirit, making one God". Since that day, the shamrock has been the emblem of Ireland, and on St. Patrick's Day the Irish like to wear a piece of shamrock to remind them of their patron saint.

St. Patrick left something else to his friends in Ireland. He composed a hymn which was first called "Cry of the Deer" but which is now often known as "St. Patrick's Breastplate". He thought that it would remind people that God's care and protection were sufficient to protect them from all their enemies. A part often sung today is the last part which is a kind of summary of the earlier verses.

Christ be with me, Christ within me,
Christ behind me, Christ before me,
Christ beside me, Christ to win me,
Christ to comfort and restore me,
Christ beneath me, Christ above me,
Christ in quiet, Christ in danger,
Christ in hearts of all that love me,
Christ in mouth of friend and stranger.

Patrick lived to be a very old man; some people say he was over a hundred years old when he died. He is buried in Downpatrick, Co. Down and many people go to his grave on St. Patrick's Day.

His flag, which is a red diagonal cross on a white ground, has been part of our national flag since the beginning of the nineteenth century.

7 ALL FOOLS' DAY

Children always get up early on All Fool's Day, for the first of April is the day for playing tricks on everybody; friends and enemies alike! No time however must be wasted, for at twelve o'clock conduct must return to normal; midday is the signal for all the jokes to stop.

It is hard to describe what kind of jokes are played, for the ingenious person will always manage to find original ones and surprise everyone. The purpose of many of these pranks is to encourage a person to do something quite unnecessary and so make a fool of himself. He must always react to the joker or else there is no fun involved. One of the commonest forms of joke is to tell a person that his shoe lace is untied and when he looks down to see, to shout at him "April Fool!" Another idea often used, is to take a letter to someone. When they open it, they find blank sheets of paper and one on which has been written: "Ever been had? April Fool!" Sending people on bogus errands is also an amusing pastime and when they return without the article, they are greeted with cries of "April Fool!" Confused messages requesting people to go somewhere or to do something which can only make them look foolish are often used.

This idea of making others look foolish is not a modern idea, for in one old book written over two hundred years ago, *Poor Robin's Almanack for 1738*, occur the lines:

No sooner doth St. All-fools morn approach,
But waggs, e'er Phebus mount his gilded coach,
In sholes assemble to employ their sense,
In sending fools to get intelligence;

One seeks hen's teeth, in farthest part of th' town;
Another pigeon's milk; a third a gown,
From strolling cobblers stall, left there by chance;
Thus lead the giddy tribe a merry dance:
And to reward them for their harmless toil,
The cobbler 'noints their limbs with stirrup oil.

There seems to be some confusion as to why we keep the first of April in this way. It seems that many Northern European countries have similar days but not always on this date.

One explanation of these strange acts is that the customs go back to the time when the kings and other important people kept jesters or "fools" at their courts and in their houses. They were as much part of the household as the servants who worked in the kitchens or on the land. Their task was to keep the master in a good humour and to distract his attention when events worried or displeased him. Sometimes they imitated the things people did or the way they spoke; if all else failed they would dance, or even perform juggling tricks. They often wore a kind of uniform, a coat or jacket of two colours with stockings of different colours. On their caps they had long points, at the end of which hung bells, which would jingle and make a cheerful noise when they moved. They often carried a stick which had a cock's or a doll's head moulded on the end of it, with a hat similar to their own. They would address remarks to this doll or pretend that it was speaking for them. Sometimes they had a balloon on the end of their stick and with this they could pretend to beat people if they were naughty.

The custom of keeping jesters has long ago died out and it is thought that the last royal jester was Muckle John, at Charles I's court, in the seventeenth century. Once a year these jesters had a holiday. There were many revels and games at the Feast of Fools; some were crude and were later forbidden by the clergy. It has been suggested that on the jester's holiday, the ordinary people could take over the playing of jokes, tricks and witticisms until noon when the real fools resumed their duties.

There is another story connected with this strange day. Some think the origin of the story is in the Far East. A queen tried to protect her son from the powers of evil which threatened to carry him away. At last the queen died and no one was left to protect the young prince. Ciongock, the evil one, appeared in a great chariot drawn by four grey dragons. When they rode across the sky, confusion spread over the earth and the plants and trees died. He challenged the young prince to fight him, but before this could begin, a goddess appeared from heaven in clouds of fire. Ciongock changed himself into many shapes to avoid her fury. The last shape was that of a great river in full flood, but unfortunately he swept the young prince away with him as well. The river emptied itself into a great lake. The villagers who lived nearby were sent to look for their prince in the lake and to ask his advice on their problems. But however hard they listened they never heard any answering voice and so concluded that they had been "sent on a fool's errand", for there was no one there.

Another myth which is also said to give rise to these April Fool tricks is that of Persephone, the goddess of Spring who was stolen away by the king of the underworld. Her mother, Demeter, heard her screams and went to find her, but it was only the echo of her voice which was heard and so she, too, had been "on a fool's errand".

Whatever the real reason behind the happenings on 1st April, it is certainly a most popular day with children. It is the day when they can be sure of making the adults look very foolish, and it is too good an opportunity to be missed!

8
EASTER DAY

The world itself keeps Easter Day,
And Easter larks are singing;
And Easter flowers are blooming gay,
And Easter buds are springing:
Alleluya, Alleluya:
The Lord of all things lives anew,
And all his works are rising too:
Hosanna in excelsis.

This is one of the many Easter carols which have been written for singing at this time. They reflect the feeling of joy and gladness which we feel, for Easter Day is the most important day in the whole year for Christians. It is a day of great happiness and rejoicing. Our churches are beautifully decorated with flowers to remind us that the earth, which has looked so dead all through the winter, is full of life and beauty. This helps us to understand the message of Easter, that death is not the end of life, but the beginning of a new kind of life. We are not sure how this day got its name. "Easter "is a very old English word, and may come from "Eastre" (Eos), or Aurora, the Greek goddess of the dawn, as the symbol of a new day.

Many churches have Easter gardens at this time. These are planted with spring flowers such as primroses and daffodils. In addition to the flowers there is a cave made from rocks: this is to represent the grave where the body of Jesus was laid after he had been taken from the cross. But in the model, the tomb in the cave is always empty for it is the resurrection of Jesus that we are celebrating on Easter Day.

Great care is taken to clean the churches carefully for Easter; the brasses and silver are polished and new candles are put ready. Then the new fire is lit from a fint and all the new candles are lit from this, including the great Paschal Candle which will burn through all services between Easter and Ascension, that is for forty days. The oil and the water for baptism are renewed at these

Easter ceremonies and thus everything starts afresh.

All Christians spend some time in Church on Easter Day, and it once was the custom to have new clothes to wear on this occasion. Easter cards are often sent at this time and are usually religious in character, but some of them have pictures of chickens popping out of eggs, representing the renewal of life.

It is the custom to give eggs as presents at Easter. Eggs are symbolic of new life and they were the commonest form of food which was eaten after the end of the period before Easter known as Lent, in which many people ate less food. In order to make them look more attractive they were hard-boiled and then dipped in a colour. Cochineal was often used as this was a harmless dye which turned the shells a crimson red. Another easy way to colour them was to tie the eggs in onion skins and then cover them with muslin and boil them. The marks on the skins produced attractive patterns.

Many games were played with these Easter eggs. "Shackling" or "Yarping" was the name given to one game that was played, where the object was to get as many eggs as possible. Egg-rolling contests were held in some parts of the country. Eggs were rolled down the sides of hills, the egg which rolled the furthest being the winner. Today Easter eggs are often made of chocolate, decorated with elaborate patterns and covered with coloured shiny paper. Some are made from cardboard and form a kind of box or container for a small present.

At this time some people eat a special kind of biscuit. These are round and flat with currants in them; in fact they are a kind of shortbread and are called Easter cakes. They should be tied with coloured ribbons in bundles of three to represent the Holy Trinity. They are given to members of the family on Easter Day.

Simnel cakes are also eaten at Easter-tide. These are made with spices, fruits and flour, and have a delicious almond paste filling and top. When the paste has been put on the top of the cake, eleven almond egg shapes are placed round the edge. These represent the Apostles, omitting Judas, the disciple who betrayed Jesus. That is why there are only eleven. Originally these cakes were made to be taken home on Mothering Sunday, which is in the middle of Lent. When children, who had left school just before Christmas, were sent away to work, they were allowed to make their first visit to their homes on Mothering Sunday and Simnel cakes were the presents which they took home with them. The cakes were then kept and first eaten on Easter Day.

Many people think that we eat Hot Cross Buns at Easter, but these spiced buns marked with a cross are made to be eaten on Good Friday. This day is part of Holy Week, the final week of the Lenten fast. The festival of Easter does not begin until Easter Day.

Special hymns have been written to be sung at the church services on Easter Day. These tell the good news of the resurrection of Jesus. One of the best-known of these has the following first verse.

Jesus Christ is risen today,
Allelujah!
Our triumphant holy day,
Allelujah!
Who did once upon the Cross,
Allelujah!
Suffer to redeem our loss.
Allelujah!

9

St. GEORGE'S DAY

St. George has been the patron saint of England since the fourteenth century. Very little is known about him, and various legends are associated with his name. The following story is one of the most commonly accepted.

He was born in the third century probably in Asia Minor, the part of the Middle East which we now call Turkey. His parents were Christians and brought him up to follow their beliefs. He joined the army and quickly rose to a high rank as he was capable of commanding groups of soldiers. He was said to be very courageous, very courteous and also very handsome.

In the course of his military duties he visited many countries, including England. In Persia, where he spent some time there are still churches dedicated to him which he may have helped to establish. Near one of them is a rose tree or group of rose bushes which is large and impressive. These are supposed to be St. George's roses, and certainly he is always associated with that flower.

In spite of his successful career in the army, he was very distressed when he heard that the Roman Emperor was ordering that all Christians should be persecuted. He felt that he could no longer be a member of the Roman army, and should confess his Christianity. He therefore gave up his position, sold all his possessions and went to see the Emperor, who on learning he was a Christian, had him tortured and beheaded.

The legends and stories about him are numerous. Many of them are about things that happened while he was on his journey to the court to see the Emperor. The best-known story is that about George and the dragon. Although the details vary in the numerous versions of the story, the main ideas are the same. It was probably added to the many other stories several hundred years after St. George's death.

St. George was travelling through a strange land when he came to a city full of strange people. They were sad and melancholy and whatever he said to them could not relieve their sorrow. After some time they told St. George that there was an extremely fierce dragon living in the marshes just outside their city. Every day it demanded at least two sheep to eat. The people supplied these, but now all their sheep were gone and the dragon demanded that children and young people should be brought instead. In order to decide who should go, lots were drawn each day. The next day the King's beautiful only daughter would have to be given to the dragon, and the King was beside himself with grief. He had begged them to spare his daughter but they had said "No! we have had to see our sons and daughters perish and the law says that the one whose name is drawn, that one must die. You have made this law for us to keep—now you must keep it yourself." So the King dressed his daughter up in her most beautiful clothes, like a bride, and sent her outside the city to wait for the dragon.

While she waited, St. George happened to come that way and saw her there. He was distressed that yet another beautiful maiden must be given to the dragon and thus perish, so he told her that he would rescue her and free the city from the power of this dreaded beast.

Calling on God to help him, St. George rushed at the dragon as it came towards him breathing out fire and smoke. He plunged his sword into the creature and badly wounded it; then he said to the King's daughter, "Give me your girdle and when I have tied him up, you may lead him into the city." When this was done the dragon followed behind her like a dog on a lead. St. George and the princess and the dragon walked back into the city, but when the people saw the terrible dragon, they ran away shouting and screaming that they would all be killed. St. George called them back and assured them that all was well, and promised to kill the dragon so that it would never menace the town again. He then taught the people about the Christian gospel. Many were baptised, including the King, and a church was founded in the city, dedicated to St. George. The King wanted to give St. George a large sum of money to show his gratitude, but he would not take anything; he asked the King to give it to the poor people of the city and he continued on his journey.

St. George's Day is 23rd April and his flag is a red cross on a white ground. There are many churches dedicated to St. George in this country; one of the best-known being St. George's Chapel at Windsor. This is the church of the Knights of the Order of the Garter. These are a small band of twenty-six people who have been specially honoured by the Queen because they have served their country in a particularly distinguished manner. They have a special uniform consisting of a mantle, surcoat and hood, with a collar of Tudor roses from which a large badge and star hangs. The badge is called the "George" because it bears the figure of St. George on it.

One modern celebration has become an annual event on the Sunday nearest to St. George's Day. This is the procession of all the Queen's Scouts in the Boy Scout movement. They parade through the streets of Windsor to St. George's Chapel where they attend a service. After this, they all march in order past the Queen (or her representative) who takes the salute. In 1952, the first public appearance of Queen Elizabeth II after the death of her father was at the St. George's Day Parade of Queen's Scouts. Thus an old ceremonial day now has a modern one linked to it.

St. George is also the patron saint of soldiers and sailors, as well as of several other countries.

10
MAY DAY

May Day Garland

I've brought you here a bunch of may!
Before your door it stands.
It's well set out, and well spread about,
By the work of our Lord's hands.
This morning is the first of May,
The primest of the year;
So ladies all, both great and small,
I wish you a joyful cheer.
Then take your Bible in your hand,
And read the scripture through;
And when the day of Judgement comes,
The Lord will remember you.
The clock's struck one! I must be gone!
No longer can I stay.
If I should live to carry again,
I'll call another May.

This is a very old May Day Carol. The word "carol" is often thought to mean a song or hymn which is sung at Christmas time, but the real meaning of carol is a song for dancing, usually in a ring, or round. In olden times 1st May was regarded as the first day of summer and a time when people felt like dancing and singing, for the dark days of winter were behind them for another year. Many of the ceremonies of May Day are very old; they go far back into the time when we had only two festivals in the year. These were the beginning of summer on the first day of May and the beginning of winter on the first of November. We shall read in a later chapter how the ceremonies to mark the return of winter have gradually been transferred to 31st October, All Hallows' Eve, but those belonging to the first day of summer are still connected with May Day. Some of these have been revived during the last hundred years and are now performed by children instead of grown-up people.

One of these is the custom of choosing a May Queen. In olden days the prettiest maiden in the village was chosen for the queen and others for her attendants, or maids of honour. In these days this custom is usually carried out in our schools. The May Queen is chosen and wears a long flowing dress if one can be found for her. Her crown is made of flowers and sometimes she carries flowers in her hands. After she is crowned there are special dances which are very old traditional dances. The most well-known is probably the Maypole dance. A pole is set up in the middle of a circle of dancers. Each one holds a long ribbon or streamer which is fastened to the top of the pole. During the dance these ribbons are plaited round the pole. Originally these were of red and white ribbons; now they are often of assorted colours.

Maypoles used to be set up on every village green and in the middle of the towns. May Day was a holiday for everyone and there were games, dances and feasting. If the young girls wanted to make themselves beautiful for this important day, it was said that they would have to get up very early in the morning and wash their faces in the dew on the grass. Those who were not chosen for the Queen or her maids would make themselves May garlands which could be carried in the May Queen's procession. These were branches of trees with flowers twined round them.

If a village had no queen, a doll might be dressed as a queen, and carried round on a May garland. Sometimes there were competitions to see who could make the most beautiful garland, and in some parts of the country these garlands were taken into the church at the end of the procession. May Day stools were also made; these were cushions fashioned out of clay, into which flowers were stuck to make patterns. These were then put at the doors of the houses for all to see and admire, and money was collected which was used for a feast in the evening.

Morris dances were a great feature of the May Day ceremonies. These were dances performed by teams of men, with bells tied below their knees and large handkerchiefs carried in their hands. A pipe and a fiddle were often played as they danced along the village streets. There were also other quaint figures who dressed up and danced round or behind the Morris dancers. One of these was called a "hobby-horse", made from a hoop with a horse's head, and from this hoop hung a cover which hid the legs of the wearer. The horse pranced round the dancers and others beat the horse with a balloon on a stick. Sometimes there were the figures of Robin Hood and Maid Marion among these dancers.

May Day was the time when the cart-horses had their parade. On ordinary days these beautiful animals did their job of pulling heavy carts round our cities. In London and Manchester, particularly, they would be specially groomed until their coats shone. Their horse-brasses would be gleaming and they would be decorated with ribbons and rosettes. Now that we have few horses this parade has been augmented, in London, by the Historic Commercial Vehicle Run, in which these old vehicles drive from London to Brighton for everyone to see. There is also an annual parade of horses in Regent's Park.

Fairs were also held in many parts of the country. They started on May Day and continued for several days. Oxen were roasted whole over fires in the streets and there were all kinds of things to buy and side-shows to visit. One of the most famous of these has given its name to a district in London which is still called Mayfair.

For the people who lived in towns there were trips into the country to go "a-maying", that is to pick branches off the May tree to decorate their houses. This is now considered an unlucky thing to do and children are persuaded not to pick May.

We have an old saying—"Ne'er cast a clout till the may is out", which means that it is foolish to leave off winter clothing until we can see the signs of summer in the blossoms on the trees.

We can still see evidence of the glorious romp that May Day used to be, although we may not always hear children singing what was once a popular verse at this time:

We come to greet you on the first of May,
And we hope you will not send us away
For we dance and sing our merry song on a
Maypole Day.

Today, May Day is celebrated in a very different way in many parts of the world, and instead of all the fun and merriment of the village green, there are military parades, and political processions, meetings of workers and young peoples' organizations. Some countries celebrate on a very big scale, and the original idea of remembering the beginning of summer is quite overlooked.

11
MIDSUMMER

Midsummer Day used to be one of the most popular days in the summer, but now, unfortunately its celebration has fallen into disuse in Britain. Many of the old customs have been transferred to May Day, which is celebrated by more people than is 24th June.

One village custom which remained longer than many others was that of having a bonfire on Midsummer Day. Bonfires have always fascinated people, for there is much that is symbolic about fire. These bonfires were started at midnight and were said to prevent evil overtaking the people. Dances were held round the fire and the young villagers were encouraged to jump through the fire for good luck. Sometimes cattle were driven through the flames and smoke in order to purify them. As the fires died down the ashes were taken up and sprinkled over the fields to make the crops grow well.

Certain flowers were said to have magic properties and were gathered at this time. St. John's wort is supposed to foretell the future and must be gathered on Midsummer's Eve. If it flourishes when brought into the house, the girl who has gathered it will be married before the year ends, but if it droops, she will not. If the plant withers and dies, that is a sign that the girl will die before the end of the year! Other plants had lucky or unlucky associations and some seeds planted at this time were supposed to grow much better than those planted earlier or later.

There were so many strange customs observed—like the one of sitting in the church porch after a long fast in order to see which spirits would come and knock at the church door foretelling deaths in the future—that it is easy to understand why strange behaviour is referred to as "midsummer madness".

In Oxford at Magdalen College, one old custom is still carried on. This is the Midsummer sermon which is preached from a stone pulpit in the

24th JUNE

quadrangle. Years ago the quadrangle was decorated with branches of trees to resemble the place in Palestine where St. John the Baptist is said to have preached. On the site where the college now stands there was once a Hospital of St. John the Baptist, thus making another link with the saint whose day falls on 24th June.

Another very old tradition which is still carried on today, and which attracts a great many visitors, is the Druids' Summer Solstice Service at Stonehenge, an ancient circle of stones dedicated to the worship of the sun, and dating back thousands of years. This circle is situated on Salisbury Plain. The service starts at dawn, for that is the time when the sun is in such a position that its rays fall on the altar stone. This cannot always be seen as it is frequently misty or cloudy at this time. But the link with the past is very firmly made as the Chief Druid says the following very old prayer during the ceremony:

God, our all-Father, permanent amid all change art Thou. Thou hast ever been and as Thou art, so shalt Thou ever be. We seek and find in Thee the glory of the dawn. We seek and find Thee when the darkness of the night has fled. The sleep of faith has ever led through night to dawn. Amen.

12
St. SWITHUN'S DAY

We can sometimes hear older people saying: "It is the fifteenth of July tomorrow—St. Swithun's Day. I wonder if it will rain? If it does, we shall have forty rainy days."

Why do they say this on this date and not on any other?

About eleven hundred years ago, in the town of Winchester, there lived a bishop called Swithun (sometimes written Swithin). The people of the town loved him very much. He was good and kind to all and there are many legendary stories told of the wonderful things he did to help his people. This is one of them.

One day a woman was going across the bridge, over the river, which led into the town. She was carrying some eggs to sell in the market. There were many other people going to market. She was jostled and pushed across the bridge, and some particularly rough men knocked her so that some of the eggs fell out of her apron in which she was carrying them. They fell to the ground and were broken. Swithun happened to be passing at the time and he picked up the eggs and gave them back to the woman. When she looked at them they were all unbroken and so she was able to sell them in the market as she had intended.

In spite of the fact that many other stories were told about the bishop's miraculous powers, he was a very humble man. He gave instructions that, when he died, he was to be buried in a simple grave, outside the cathedral. It was the custom at that time to make very elaborate tombs inside the cathedrals to bury kings, bishops and other important people. Swithun wished his grave to be in the ground which lay round the cathedral because he thought that the sky made a better roof than any grand tomb made out of stone.

So when he died, the people of Winchester did as he had asked them and many came to visit his simple grave, thinking that the powers of their bishop would still be able to help them in their troubles.

Several years later, the Pope gave permission for Swithun to be called Saint Swithun. The monks who lived in Winchester were very concerned that the saint's grave was such a humble one and they decided that it should be removed and put inside the cathedral. The date they fixed for this was 15th July, and a grand procession and service were arranged to take place on that day. Unfortunately when the day came, instead of the fine day for which they had hoped, it rained. The old accounts of this tell us that such rain had never been seen before and so the procession had to be put off. The weather remained very bad; in fact it rained very hard for forty days. At last, the monks became alarmed and said that God could not be pleased at what they had decided to do and was trying to prevent them from carrying out their plan. Instead, they built a little chapel over the grave and many people came to worship in this shrine.

After this, on 15th July, people have said that if it rains on that day, it will rain for forty days, just like it did on the first St. Swithun's day.

13
ALL HALLOWS' EVE

All Hallows' Eve, 31st October, was the most terrifying night in the whole year. This was thought to be the night when witches held revels and parties and cast spells over the people which would harm them. It was also the last night of summer, the harvest was gathered in and food was secure for the months ahead. The land was now bare and empty, and November was known as the "month of the dead". The nights grew longer and so the hours of daylight were less. The winter storms arrived and gales blew. Often trees were blown down and houses were damaged. Then came the snow and ice of the winter; the long, cold time when the people found it difficult to visit their friends if the paths were blocked by snow. It was also very hard work to find food and to attend to their animals. If these died, the people would be very poor indeed and might even starve. There was also the fear of illness and death. So Hallowe'en was a sad time for many people. They were frightened of the witches and of their powers to hurt people by means of spells.

Many of the customs of Hallowe'en are very old indeed. We now keep them in the form of games but they were originally used to stop the witches from doing harm. Apples and nuts appear in many of the games. To us these are delicious but quite ordinary items of food. But to primitive people they meant something more. The hazel-nut was a symbol of wisdom and anyone who wished to become wise was advised, not only to eat the nuts, but also to treat the tree with great respect. Apples were supposed to be the symbol of life and eternal happiness. These could not only be eaten, but from them a drink could be made which had power to make men forget their troubles and think themselves wiser and braver than they really were.

One game called "Bob Apple" consists of trying to bite an apple floating in a tub of water. Sometimes the hands are tied behind the back to make it harder. If the player succeeds in taking a bite, he is allowed to eat the apple and can be sure of good health and happiness for the coming year. There are different versions of this game which are played in various parts of the country.

One of the commonest games with nuts is to place two nuts side by side in or near the fire. These are given names. If the nuts burn quietly and steadily then this foretells marriage, but if the nuts crack and pop and fly in all directions, this denotes a quarrel.

These old games seem to have two purposes; one to ensure that health and happiness will come to the player, perhaps by removing any chance of a spell from a witch, and the other is to foretell the future. Carrying salt, throwing it over one's shoulder or putting it in the keyhole were considered good ways of keeping witches away.

A young girl was advised to peel an apple, taking care to keep the peel unbroken. She should then wave the skin three times round her head and let it fall. It would then form the initial letter of the name of her future husband. Another way for a girl to discover whom she would marry, was to brush her hair three times in front of a mirror at midnight and she would see the face of her future husband peering over her shoulder.

Lanterns placed in windows or carried in the hand were also used to frighten witches. These were made by scraping out the inside of pumpkins, swedes or turnips and cutting holes for eyes and mouth. A candle was then put inside. These lanterns are also now used for decoration and for frightening people instead of witches!

In country areas, hundreds of years ago, bonfires were lit round villages and fields. These were supposed to give more power to the sun which was getting weak at this time of the year and also to purify the fields after the crops had been gathered. These fires would also serve to keep the witches away from the village.

Now most of these old customs are only remembered in the form of games. It is quite common to hold Hallowe'en parties; to dress up and wear masks, to play these games, to decorate houses with lanterns and to hang old charms on the walls. In this way fun is made of customs which were once used to dispel the fear of witches and of death.

14

GUY FAWKES' DAY

Guy Fawkes' Day on 5th November is celebrated each year by lighting bonfires and setting off fireworks. This has been done since 1605. This was the year in which Guy Fawkes was discovered in the act of trying to blow up the Houses of Parliament together with all the members of Parliament and the King. There are many stories in the history of our country in which citizens have planned and plotted to kill the reigning king or queen, but they are nearly all forgotten. However, no one ever forgets Guy Fawkes and his escapade, and in the following story you will see why,

When Queen Elizabeth I died there was no direct heir to the throne. This was quite a problem at the time, but the nearest person descended from Henry VII, was James VI who was then King of Scotland. He was invited to become King of England—James I— and thus to unite Scotland with England and rule over both countries. This was not a popular move with many people, especially with those who were Roman Catholics, for James I had been brought up as a Protestant and did not encourage Roman Catholics. It was very difficult for their priests to carry out their duties and many went about in terror for their lives. In large houses, secret rooms were made for them so that they could be safe from persecution. These were called "priests' holes" and many lives were saved in this way.

James was not a popular king, and many people were unhappy and uneasy under his rule. He quarrelled with members of Parliament because he was not consistent in dealing with matters concerning this country's relations with foreign powers, and was too severe in his treatment of Roman Catholics.

Eventually, two years after he became king, a group of friends decided that matters were so seriously wrong that something must be done to stop them from getting worse. Robert Catesby and his friends planned a very daring solution to this problem. They would destroy the king and all the members of Parliament at one stroke. This could be done by exploding gunpowder under the Palace of Westminster, which is now called the Houses of Parliament, on the day when the king was opening Parliament. By choosing this day, they would be sure of all members being present at one time. Then the country would be free to make a fresh start.

There were difficulties to be overcome before this plot could be carried out. One of these was the difficulty of getting into the Houses of Parliament unobserved. This was solved by one of the group of friends renting the house which adjoined the House of Lords. From the cellar of this house they would be able to enter the cellar of the House of Lords, and it was here that they planned to plant their explosives.

The second difficulty was to find someone who was an expert in the field of explosives who could direct the operation. Several members of this group of friends came from Yorkshire and they remembered their old school friend, Guy Fawkes, who was a soldier. He had spent ten years fighting for the Spaniards in Holland and the Low Countries, and knew all about the use of explosives. There was an added advantage in inviting Guy Fawkes to join their group, for he was not known in London and he would therefore be free to come and go as he pleased without anyone connecting him with Robert Catesby's friends.

So the work began. A way was made into the cellar, and thirty-six barrels of gunpowder were brought and stored there. Some people think the barrels were brought over the river Thames in boats, until there was a very large store amounting to nearly two tons underneath the hall where the opening ceremony was to take place. Some accounts of this story tell us that the barrels of gunpowder were covered with coal and iron bars which would give more force to the explosion. Then the whole pile was covered with firewood so that anyone looking into the cellar would not suspect anything. Finally Guy Fawkes laid a trail to the gunpowder which would burn for about a quarter of an hour thus giving him time to get away, after lighting it, before the explosion.

The date of the opening of Parliament was altered several times and Fawkes had time to go back to Flanders and return before the date was finally decided. This meant that the cellar had to be inspected to see if the gunpowder had been ruined by the damp and to replace any if necessary. This delay greatly increased the danger, but, at last, the 5th November was chosen.

Guy Fawkes went down into the cellar on the evening of the 4th to begin a long and lonely wait until the hour of the opening ceremony, when he would perform his deadly task. But the plan had been betrayed and the king's men searched the cellars and discovered the gunpowder and Guy Fawkes ready to ignite it. He was arrested and taken straight to the king's bedroom where important ministers had been summoned, although it was one o'clock in the morning. He refused to give the names of his fellow conspirators and so was imprisoned. It was not until he was put under torture several days later, that the names of the group were disclosed. They were finally all executed in January 1606.

Parliament was so relieved that the explosion had been averted, that the Church was ordered to say prayers of thanksgiving on every following fifth of November. Thus the day became one of celebration. As the years went by, it became popular to have bonfires as part of the celebrations and the custom of "burning the guy" became established. This guy was a figure made out of rags to represent Guy Fawkes and it was then put on the top of the bonfire. Fireworks were also used, and these celebrations became very

wild indeed. Blazing barrels of tar made village streets dangerous places to be in on Guy Fawkes' Night.

At last regulations were made declaring it illegal to let off fireworks in the street or to burn barrels of tar. Now people may have fireworks in their own gardens or public places where permission has been given for this purpose. In Brighton, the beach may be used for letting off fireworks and making bonfires on 5th November. Some villages in Sussex, for example, collect rubbish and prepare a huge bonfire on the village green and hold grand displays of fireworks, making it into a gala night, in which many visitors join. In Lewes, there is usually a procession through the town at the end of which figures of the Pope as well as Guy Fawkes are burnt.

It is a common sight to see children carrying or pushing in a pram or cart a home-made figure of rags or sacks and calling at the same time, "A penny for the guy". Fireworks are usually bought with the money that is collected.

This is a very popular festival and one which is likely to continue for many years. Should we ever be in danger of forgetting it, there is always the traditional rhyme to remind us.

Please to remember
The fifth of November,
With gunpowder, treason and plot.
I see no reason
Why gunpowder treason
Should ever be forgot.

Perhaps we should add that this account of the "Gunpowder Plot" is only one story of several, and some people doubt whether there was a plot at all!

15

St. ANDREW'S DAY

The Festival of St. Andrew is kept on 30 November.

St. Andrew has a special place among our patron saints, for he is the only one to have been living in Galilee at the same time as Jesus. He was a fisherman and the brother of Simon Peter. When Jesus selected his small group of friends to live and work with him, these two brothers were the first ones to be chosen. We are also told that it was Andrew who found the boy whose loaves and fishes fed the great multitude of people who came to hear Jesus preach.

We know very little about his later life. After the resurrection of Jesus, when the disciples were scattered over many countries, teaching and preaching to the people, Andrew is reported to have visited different parts of the Middle East. He preached in Syria, Turkey, farther north on the shores of the Black Sea and in what is now Russia. He was a favourite saint of the Russian Church, and became the patron saint of Russia. He founded many churches and converted many people to Christianity.

At last, in one province he angered the Roman proconsul whose wife had been influenced by his preaching. She became a Christian and wished her husband and all her friends to do the same. The Proconsul was so angry that he ordered Andrew to be killed in order to free that part of the world from this new religion. He decreed that Andrew should be crucified, but not in the same way as Jesus, nailed to an upright cross. He would be bound to a cross made in the shape of a letter X as this form of death was slow and lingering. Even in these ghastly circumstances Andrew was able to meet his death with courage and fortitude. The story tells us that in the two days in which he was gradually dying, he preached to twenty thousand people from the cross and made many converts.

Many years later, when Christianity was the

official religion of the Roman Empire, his body was taken to Constantinople by the Emperor's orders and buried there for some time. A man called Regulus was in charge of the grave and he had a strange dream. It was so compelling that he was sure it was a message from an angel. In the dream a voice told him to take the body of St. Andrew and he would be guided to find a safe resting place for it. At last after many years of wandering he came to Scotland and buried the relics in a spot which became known as St. Andrews. There is a town of this name on the East coast of Scotland, and this is believed by many people to be the spot where St. Andrew's grave was finally located.

About the middle of the 8th century Andrew became the patron saint of Scotland. His flag is known as St. Andrew's Cross. It is a white diagonal cross on a dark blue background. This cross can be seen in the Union Jack.

16 CHRISTMAS DAY

Christmas Day, 25th December, is the most popular day in the whole year, not only because it is the day on which presents are given and received. We decorate our churches, houses and shops, and in some towns even the streets have coloured lights hung in decorative patterns which look very attractive. The day we call Christmas Day is only part of the festival of Christmas which stretches over several days, lasting until "twelfth night" the eve of Epiphany (5th January) representing the day Christmas used to be celebrated before the change of the calendar in 1752. In some parts of the world it is still celebrated on that day.

Long before Christianity was brought to this country, there had been a day of celebration towards the end of December. The hours of daylight become less and less up to 21st December which is the shortest day in the year, and after this, the days become longer by a very gradual process. In very ancient times, people used to light large bonfires about this time to help the dying sun to retain its warmth and not get any weaker. This practice developed into a kind of festival to the sun and the sun god. It was thought that by performing certain rites the sun god would be pleased and would shine more strongly on the earth.

In Roman times this festival was called Saturnalia, in honour of the corn-god Saturn, and feasting and merry-making went on for seven days, during which time presents—often wax candles and clay dolls—were exchanged between families and friends. To the early Christians it seemed natural to link this festival to their own celebrations in honour of the birth of Jesus Christ, God's son, and so Saturnalia became Christmas (the mass, or festival, of Christ), and was finally fixed as 25th December.

As well as sun worship at this time of the year, attention was also paid to the trees of the woods and forests. We have already heard how people thought the apple and the hazel nut trees had special properties; they also thought that the evergreen trees were magic because they kept their leaves all through the long hard winter. The holly tree, with its bright red berries, was popular because at a time when no flowers were growing, these berries made a vivid splash of colour in the greens and browns of the forests. Branches of the holly tree could be cut and taken into the houses where they lasted for some time. Ivy leaves were also used for decoration because their long trails could be twined in all directions and were easy to hang on the walls.

The worship of the sun and the trees had gone on for hundreds of years before the first missionaries brought Christianity to Northern Europe. These early Christians thought the people were very wicked to worship the sun and the trees, but realised that they would not give up these ceremonies unless they had other things to put in their places. Gradually as had happened in Southern Europe, the Christian festival of the birth of Jesus was introduced in order to combine some of the better elements of the old ceremonies with Christian ideas.

In some of the things we do at Christmas we can see how they might have been part of the earlier ceremonies. For instance, we still decorate our churches and houses with evergreens—holly and ivy—and these trees are often mentioned in carols which are sung at this time. In some of these carols, the parts of the tree are linked with Mary and Jesus to emphasise the Christian aspect of the festival. This is one of these very ancient carols:

The holly and the ivy,
When they are both full grown,
Of all the trees that are in the wood,
The holly bears the crown.

The rising of the sun
And the running of the deer

The playing of the merry organ,
Sweet singing in the choir.

The holly bears a blossom,
As white as the lily flower,
And Mary bore sweet Jesus Christ,
To be our sweet Saviour:

The holly bears a berry,
As red as any blood,
And Mary bore sweet Jesus Christ
To do poor sinners good:

The holly bears a prickle,
As sharp as any thorn,
And Mary bore sweet Jesus Christ
On Christmas day in the morn:

The holly bears a bark,
As bitter as any gall,
And Mary bore sweet Jesus Christ
For to redeem us all:

The holly and the ivy,
When they are both full grown,
Of all the trees that are in the wood,
The holly bears the crown.

The only evergreen which was never taken into churches was the mistletoe. This is a strange plant which cannot grow on its own but must fasten itself on to another plant or tree. Its berries and their juice were supposed to have special powers (they are, in fact, poisonous if eaten) and they were used as magical emblems. There are many legends about people being killed or wounded by this plant, and the churches banned its use. Now we only use it in houses, where it is hung over a doorway or from a light and it is thought to be lucky to be kissed "under the mistletoe".

The Yule log was another very old custom. A fine log was selected in the forest and then decorated with ribbons. This was then dragged home and anyone who could do so joined in a procession, singing and dancing as they brought back the huge log to burn in the open fireplace all through the days of Christmas. This is a custom which is practically never seen today, now that open fires are disappearing from our houses.

Traditional games were also played at this time. Some of these required people to dress up and are perhaps the fore-runners of the Nativity plays which are acted in churches and schools. Some games were played after Christmas Day and kept up the Christmas spirit until the New Year. One of these games which was played on Twelfth Night has come down to us in the form of a carol, and seems rather strange until one realises that it was meant to be a game. One person started it, the second continued and the third repeated it and so on. Everything had to be said or sung in the correct order or a forfeit had to be paid. This is the song or carol:

The first day of Christmas, my true love sent to me
a partridge in a pear tree.

Each day another gift is added, and so the verses get longer and longer, and more and more difficult to remember in the right order!

On the second day of Christmas, it is
two turtle doves and a partridge in a pear tree.
On the third day,
three French hens;
on the fourth day,
four colly birds (*black birds*);
on the fifth day,
five gold rings;
on the sixth day,
six geese a-laying;
on the seventh day,
seven swans a-swimming;
on the eighth day,
eight maids a-milking;
on the ninth day,

nine drummers drumming;
on the tenth day,
ten pipers piping;
on the eleventh day,
eleven ladies dancing;
and finally, on the twelfth day—
The twelfth day of Christmas, my true love sent to me
Twelve lords a-leaping, eleven ladies dancing,
Ten pipers piping, nine drummers drumming,
Eight maids a-milking, seven swans a-swimming,
Six geese a-laying, five gold rings,
Four colly birds, three French hens,
Two turtle doves, and
A partridge in a pear tree.

Present-giving at this time has been popular from very early days, although at first, presents were very simple things. It was also a pre-Christian custom for the rich lords and chiefs to wait upon and serve the servants at the feasts which were held. This custom can still be seen in many forms today at Christmas parties which are held in works, factories, shops and many more institutions. The managers and those in charge wait on the workers and provide them with Christmas food and entertainment.

The special food which was eaten at Christmas consisted originally of food which could be obtained in this country. Some of the early carols sing the praises of the boar's head which was a very early type of Christmas fare. It was carried into a great hall, where many people were gathered together to eat this delicacy, which was roasted and then placed on a great platter with an orange or lemon in its mouth. Roast beef, geese, capons, peacocks, and swans were other things which were eaten by the rich people. Later, in the time of Queen Elizabeth I, turkeys became very popular as a Christmas dish and have remained popular ever since.

Today we eat mince-pies which have dried fruits

as their main ingredients, but these were not always sweet pies. They were first called "mutton pies" or "shred pies" from the suet they contained and they were filled with small pieces of meat mixed with fat in the form of suet. As meat became scarce, other things were substituted and today we have "mincemeat" which is very far removed from any normal meat!

The first Christmas puddings were made from a kind of soup. This was obtained by boiling the bones of beef and mutton and making this thick by adding bread and any dried fruit which was available, such as raisins, and seasoning it with spices. This was eaten with the Christmas meat or poultry. It was many years later that the version of pudding we know became popular. The ordinary people had more fruit and nuts than meat, and these were used and the puddings boiled for a long time until they were solid. They were then set alight with burning spirit and carried to the table amid cheers and songs.

Another feature of our Christmas celebrations is the custom of bringing a fir tree into the house and decorating it with coloured lights, and presents. This is a very old custom in Northern Europe, but it was popularised here by Prince Albert, the German husband of Queen Victoria, in the middle of the nineteenth century. Photographs were taken and drawings were made of the royal tree and this idea was copied. Now almost every family has a tree and many are placed in the windows for all to see. Electric lights have taken the place of candles and many decorations can now be bought in the shops instead of being home-made.

No Christmas would be complete without the figure of Father Christmas. According to the legend he has a red robe trimmed with white fur, rides on a sleigh drawn by reindeer and visits all the houses where there are children. He comes after the children are asleep, enters the house by way of the chimney and fills the stockings (previously hung up by the beds or the fireplace) with toys. This idea of Father Christmas, or Santa Claus as he is sometimes called, is based on the story of St. Nicholas whose feast is on 6th December. On this day he was supposed to distribute gifts to children who had been good, and sometimes he was accompanied by someone who would beat the naughty children.

Just over one hundred years ago the first Christmas card was designed. Cards became popular when the pre-paid penny post was introduced, and people could buy stamps to stick on cards and envelopes. At first designs were comparatively simple, but later they were decorated in colour, with lace borders and sometimes silks and velvets were used. Sometimes sprays of holly, ivy or mistletoe were drawn on the cards and then there appeared drawings of the robin. This is one of the birds which can be seen all through the winter, and is greatly loved by the British people.

It is possible to see early Christmas cards in the Victoria and Albert Museum in London. In early Victorian days there were not very many cards sent as they were not easy or cheap to make, but now there are millions of cards sent every year, and the designs have become very diverse, many of them having nothing to do with the real meaning of Christmas.

Since the end of the Second World War a new Christmas custom has appeared in our country. In the middle of December a very large Christmas tree is sent from Norway. It is chosen for its height and beauty, and it is decorated with coloured lights and placed in Trafalgar Square, in the heart of London. The Norwegian people fought as our allies during the war and they send us this gift as a symbol of friendship. Carols are sung round the tree each night by choirs from many different parts of London and everyone is free to join in the singing if they wish.

Sometimes parties of people go from house to house singing carols in order to collect money. These are called "Waits" (from an Old French word meaning "to watch").

Sometimes there are Christmas trees set up in public places, like railway stations, and choirs gather there to sing carols and to collect money for charities. Christmas is the time of the year when we like to feel that the poor people are thought of and cared for.

If you have found these chapters interesting, you may want to read some books which will tell you more about customs in Great Britain.
Here are some titles:

English Customs and Usage
Christina Hole *Batsford*

Scottish Pageantry
Albert Makie *Hutchinson*

Folklore of the British Isles
E. Hull *Methuen*

The Twelve Days of Christmas
M. and J. Hadfield *Cassell*

Christmas Customs and Folklore
M. Baker *Shire Publications*

Compliments of the Season
Ellinger and Holloway *Penguin Books*

Discovering English Fairs
Shire Publications

Your Public Library will help you to find books which describe particular customs in your locality. There is also a book called, *Royal Pageantry, Customs and Festivals of Great Britain and Northern Ireland,* by *Purnell.* This describes where and when you can see these events to-day.

The Year Book by William Hone and *The Every-Day Book* by the same author were written over a hundred years ago. These books have something to say about almost every day in the year and are very quaint and fascinating to read. It is sometimes difficult to find these books, but, again, the Librarian will be able to help you in your search for further information about days of the year in Great Britain.